Dedicated to:

Expanding children's answer to
"What will you be when you grow up?"

IVY LEAGUE UNIVERSITY
presents
AKIMIE
The degree of
Doctor of Astrophysics

First Printing, 2016

ISBN 978-0-9981541-0-7

www.TheScienceStarters.org

Thanks to:

Dr. Aomawa Shields, Astronomer & Astrobiologist,
University of California, Los Angeles

Dr. Debra Fischer, Astrophysicist, Yale

Dr. Kathryn Hadley, Astrophysicist,
Oregon State University

Table of Contents:

Andromeda in ultraviolet, captured by NASA's Swift satellite

Hello everyone.
Thank you for coming to see me,
Astrophysicist Akimie.

Tonight we will journey far and wide
To explore what the universe holds inside.

Our adventure begins where our
naked eye can see,
On the ground beneath you and me.

Then we'll turn our eyes to the sky
to learn what's in it and why.

Down we go, to the earth beneath our feet.
I hope you're ready to learn
something neat!

As you sit still, the world spins around.
It rotates over five million feet an hour,
like a giant merry-go-round.

Each rotation of our
wonderful world ride

Entomologist

FIELD
NOTES

5

Ecologist

Brings a new day for
us to explore outside.

During the day and at night,
Your eyes can leave this world
and take you on a magical flight.

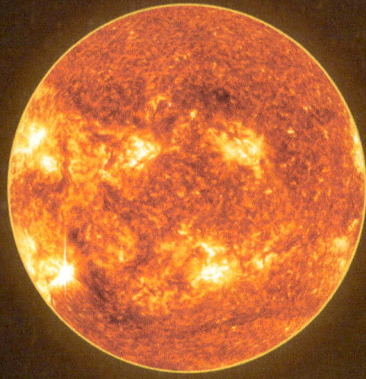

When the sky is clear and blue,
you can see a single star.

It's almost 100 million miles away
and your eyes see it from that far.

Our sun is the daytime star that you see.

It's so massive a million Earths
could fit inside easily.

Mercury Venus EARTH Mars Jupiter Saturn Uranus Neptune

Our Earth is attracted
to and orbits around the sun,

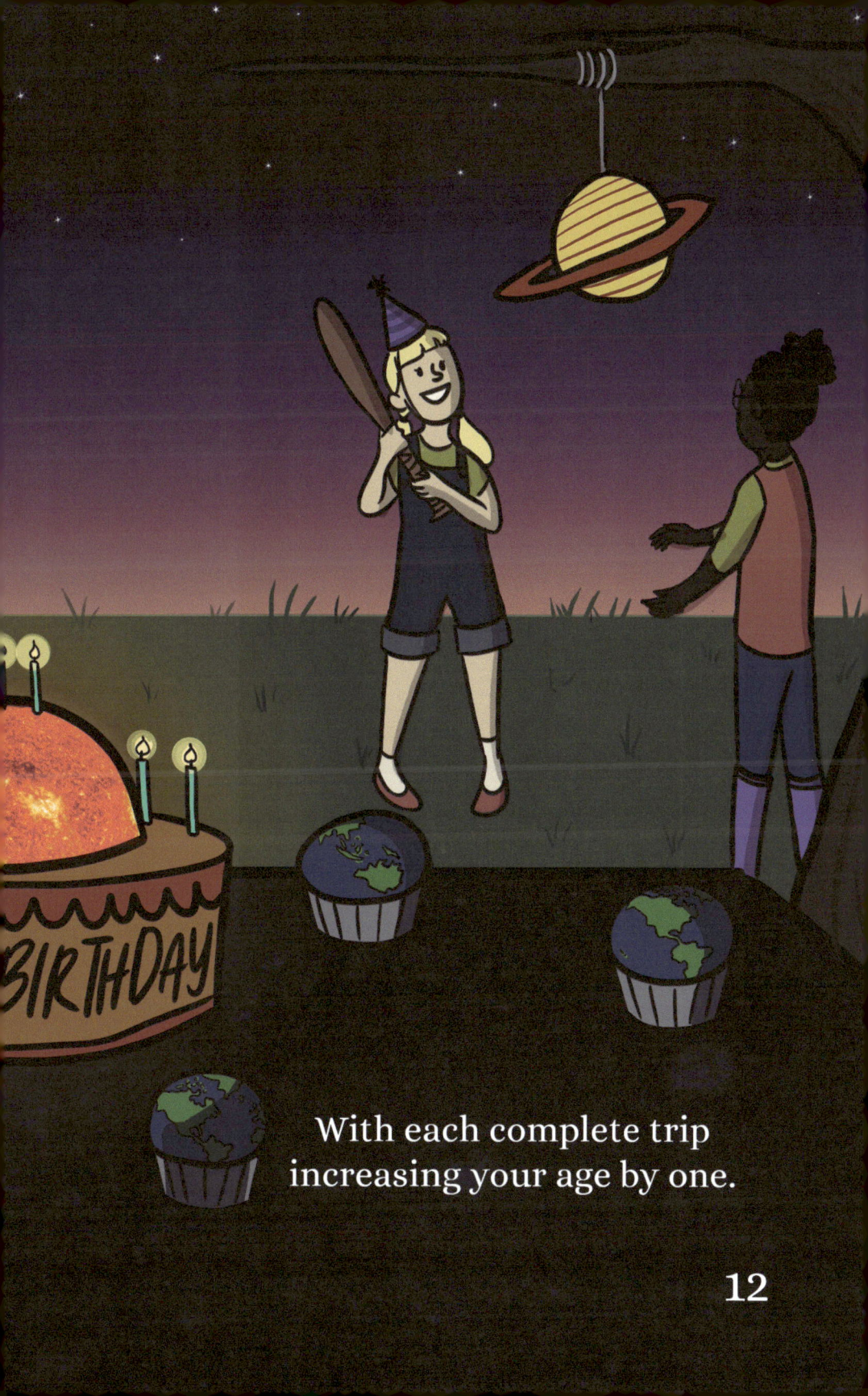

With each complete trip
increasing your age by one.

As night approaches, the sun will set,
and the moon may rise,
Ushering in darkness and surprise.

DISTANCE LIGHT TRAVELS

STAR

EARTH

S/186,000 miles

Yr/31,536,000 s

MILES FROM
EARTH TO STAR

MILES LIGHT
TRAVELS IN 1
SECOND

X

NUMBER OF
SECONDS
IN A YEAR

earth to Sirius A:
50.5 trillion miles

HOW MANY YEARS IT TAKES
FOR LIGHT FROM STAR
TO REACH EARTH

The darkness of night allows us
to see into the past

Because the light you see from distant
stars can only travel so fast.

Sirius A is the star with the brightest glow.

It's so far away that you see it as it was
almost nine years ago.

14

I used to look into the night sky and wonder,
Do stars come from a reaction between
lightning and thunder?

While studying I learned my hypothesis
wasn't right.

Stars come from star nurseries,
which give them their light.

I'll show you a star nursery,
but first we must equip our eyes.

With the help of a telescope,
I'll reveal the surprise.

Westerlund 2, captured by the Hubble Space Telescope

Using the telescope named after
Astronomer Hubble,

We've confirmed that our universe
isn't a tiny bubble.

It's black space where matter is strewn,
And it's expanding just like a balloon.

Let's use the Hubble Space Telescope to see
what else we can find.
By doing so, we'll expand our minds.

I see something
big and bright and long and thin.
It's a spiral galaxy moving in a sideways spin.

If we continue to look, we'll see billions more.
Ours is the Milky Way galaxy,
and Andromeda is right next door.

Within each galaxy are billions of moons, planets, stars, and black holes.

Understanding why all of this exists and how it functions is astrophysics' primary goal.

To answer these questions we
must travel back through time.

We'll look back roughly 14 billion years to
understand the Big Bang paradigm.

Everything in our universe was trapped
in a kind of vault,
And this vault was much smaller than
a single crystal of salt.

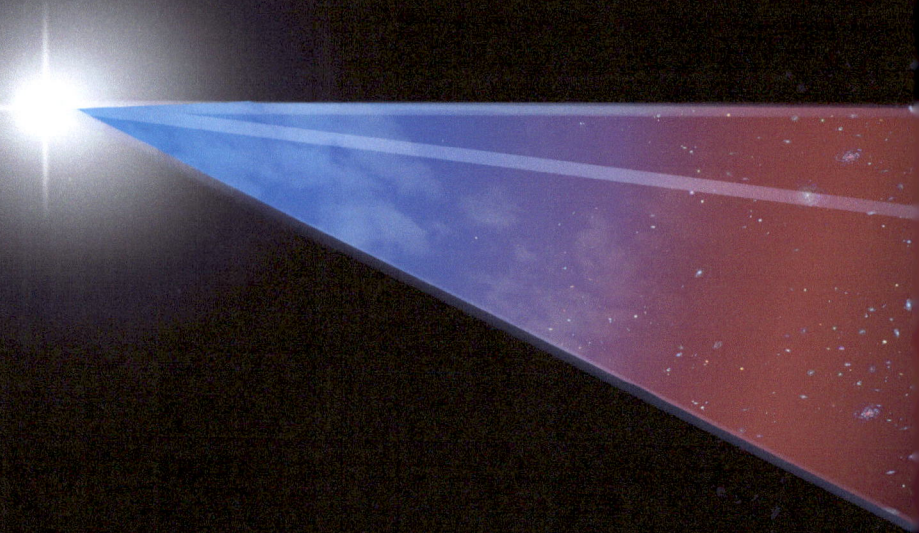

13.8 billion years ago the vault
began to shatter
Releasing space, time, energy, and matter.

The baby universe was very dark and hot.
As it aged and cooled, it changed a lot.

The darkness became speckled with light
once the first stars emerged
Because within each star, nuclear fusion surged.

Eventually, the first stars died and exploded,
Filling the universe with the stardust
of which all life is coded.

28

This stardust links the present with the past.
How can this be? I'm happy you asked!

We've learned that our universe must recycle,
And the oldest stardust exists in everything,
including galaxies, people, and my puppy, Michael.

My time is up, but I have a few more things to say.
I'm proud of you for learning something new today.

A few minutes ago your knowledge of
the universe was much smaller
And now, you're an astrophysics scholar!

If you'd like to know more and perform
some research of your own,

You may end up becoming
THE GREATEST ASTROPHYSICIST
science has ever known!

The Science Behind the Rhymes

What do you mean when you say "naked eye?"
The term naked eye means observing your surroundings without the help of technology, like telescopes.

Why does Earth rotate like a merry-go-round?
Imagine an ice skater spinning with her arms stretched out. As she pulls her arms toward her chest, she spins faster. As she stretches her arms out, she slows, but she doesn't stop. To stop her from spinning, force must be applied. If no force is applied, she'll continue to spin. This same concept applies to Earth and its rotation, and to our solar system in general. Our entire solar system was created within the rotating disk of the Milky Way galaxy. Because the matter within the disk was rotating when our sun and planets were formed, they rotate too.

How come the light I see from Sirius A is almost 9 years old?
Light has a constant speed, and it's the fastest-moving thing that we know of. Using the distance to Sirius A and the speed of light, we can calculate how long ago the light we see from Sirius A left its surface.

$$(50.5 \text{ trillion miles}) \times (\text{second} / 186{,}000 \text{ miles}) \times (\text{year} / 31{,}536{,}000 \text{ seconds})$$
$$= 8.61 \text{ years}$$

What do you mean when you say the Earth is attracted to and orbits around the Sun?
The Earth and Sun experience a gravitational attraction. In other words, all massive objects are attracted to each other through the force of gravity.

Hole in our sun, captured by NASA

The Science Behind the Rhymes, continued

What is a hypothesis?
A hypothesis is a testable prediction. If the prediction is not testable, it's not a hypothesis.

How do stars form?
Star formation occurs within stellar nurseries, which are pieces of giant molecular clouds. Overwhelming gravitational force within these structures causes a large cloud of dust and gas to condense into a disk. As time goes on, forces within the disk cause it to decrease in size and increase in density and temperature. Once the temperature reaches roughly 18 million degrees Fahrenheit, nuclear fusion starts, and the star begins to shine.

What is a telescope?
A telescope is a tool that allows us to see things in space that the naked eye cannot. Telescopes are typically long tubes containing multiple glass lenses, similar to magnifying glasses. These lenses concentrate light and magnify objects.

The first person to ever point a telescope at the night sky was Astronomer Galileo, and he did so in the 1600s.

What is the Hubble Space Telescope?
The Hubble Space Telescope is a telescope designed to detect light within the visible spectrum—in other words, light that our eyes can see.

The Hubble Space Telescope is the first space-based telescope, meaning it is the first telescope to live in space.

What is an astronomer?
A scientist that studies the stars and other objects in space.

Who is Astronomer Hubble?
Edwin Hubble is one of the scientists responsible for expanding our view of the universe. He proved that galaxies other than our own exist and went on to help develop the theory of the expanding universe.

What do you mean when you say the universe is expanding like a balloon?
Imagine drawing dots on a deflated balloon and then blowing the balloon up. As the balloon gets bigger, the space between the dots increases. This same concept applies to our universe and how it's expanding, except the balloon is spacetime, which is the fabric of the universe, and the dots are galaxies. Spacetime is expanding at roughly 240 feet per second per million light-years.

What is a galaxy?
A galaxy is a collection of celestial matter held together by gravity. Celestial matter is dust, gas, and objects like stars, planets, moons, and black holes.

What is a black hole?
A black hole is a gravitational trap that even light, the fastest thing in the universe, cannot escape.

What do you mean when you say the Big Bang paradigm?
A scientific paradigm is a framework of thought structured by evidence collected using the scientific method.

Today's scientific paradigm, or prevailing framework of thought, for the creation of the universe is known as the Big Bang theory.

35

What was the first generation of stars like, and how was it different from the following generations?

When the first generation of stars formed, the only elements that existed to make them were hydrogen and helium.

When the first stars died, they released heavier elements into the universe, like iron and oxygen, which became part of the starting material for future generations of stars.

The heavy elements released by the first generation of stars were created by nuclear fusion.

What is nuclear fusion?

Nuclear fusion is when atoms unite, producing a single, different atom. For example: If two hydrogen atoms fuse with two neutrons, a single helium atom is produced.

Nuclear fusion of elements up to iron releases stupendous amounts of energy in the form of light. The fusion of iron and heavier elements consumes energy. The energy required for these reactions is produced when stars explode.

Why does a star explode when it dies?

A star explosion is triggered by a sequence of nuclear fusion reactions and gravitational collapses, followed by fragmentation of the star's core by light.

Hydrogen is the first element to undergo nuclear fusion. Once all of the hydrogen has fused into helium, the core of the star will give way to gravity and collapse into a denser, hotter core. Then, helium will fuse into carbon. Once all of the helium has fused into carbon, the core of the star will collapse again, getting even denser and hotter. This chain reaction continues until the core of the star is blazing hot iron. At this temperature, light has enough energy to pierce the iron atoms, breaking them into fragments. As fragmentation occurs, the core collapses into a small object at the centermost point of the star. Upon hitting the center, the matter rebounds outward, creating a massive explosion.

The Science Behind the Rhymes, continued

What do you mean when you say "stardust"?

Stardust is the ash of exploding stars. This ash contains heavy elements like iron, oxygen, and carbon. Iron is in the Earth's core, and it's also in our blood. Oxygen is in water and air. Carbon is found in all forms of life. These elements were created deep within the first generation of stars, via nuclear fusion, and released into the universe as dust when they exploded.

How do scientists know that the Big Bang actually happened?

In the first half of the 20th century, a team of physicists hypothesized that our universe was produced by a huge burst of energy billions of years ago, and that leftover heat from the explosion would exist as microwave radiation throughout space. In 1964, two radio astronomers accidentally discovered the microwave radiation predicted by the Big Bang theory. This discovery provided the first set of data in support of the Big Bang theory. Since the 1960s, scientists have continued discovering overwhelming amounts of data that support the Big Bang theory such as the expansion rate of the universe, which if put in reverse, says that the universe was created roughly 14 billion years ago.

How do scientists know the universe likes to recycle and that recycled stardust exists in everything?

Albert Einstein discovered the Law of Conservation of Mass-Energy, which you may know as $E = mc^2$. Simply put, the universe says that nothing can be created or destroyed, but it can change form.

Imagine a star fusing helium into carbon. The helium used to make the carbon came from the fusion of two hydrogen atoms, and the hydrogen atoms came from the Big Bang. Now imagine a baby tree growing. The backbone of life, regardless of its form, is carbon. It's the fundamental building block of life's DNA. Where did the carbon come from? It came from the carbon makers ... the stars.

Becoming an Astrophysicist

What is an astrophysicist's typical day like?
Astrophysicists spend their time thinking, reading, writing, and doing math. These tasks are tools they use to analyze data obtained from trips to observatories and to write papers explaining what they've learned. Experienced astrophysicists may also teach classes, manage teams of researchers, and help develop future plans for NASA. The everyday efforts of astrophysicists are aimed at answering big questions like "How did the universe get here?" and "What is the universe made of?"

If I want to be an astrophysicist, how do I become one?
Below are a few guidelines to help you on your journey:

- Learn and develop a passion for math, the language of the universe. Learning a new language is challenging, but you learned one, which means you can learn another.

- Take advanced math and physics classes in junior high and high school. If this isn't an option for you, do your best in the classes you can take.

- Be an active participant in your local astronomy society. If there isn't one, you can create one.

- Regularly visit NASA's website to keep current with space news.

- Receive a bachelor of science degree preferably in math, physics, or astronomy.

- Keep in mind that your professors are one of your greatest resources. They may seem intimidating, but don't let that stop you from developing a professional relationship with them. Attend their office hours, ask them questions, learn about the path they took to become who they are today, and ask for advice.

- Work as a research assistant in a lab studying things that interest you while obtaining your bachelor's degree.

- Obtain a PhD in astrophysics.

"For students who want to be an astrophysicist, the most important thing is to not give up. There will be days when having courage means just making it through your classes and exams. Instead of spending time thinking about your grades, spend time thinking about what you're learning. Be up front about what you don't know - that's the only way to learn. Study things that interest you even if it seems like a digression from your main path. Be curious and inquisitive. Ask questions instead of asserting truths."

– Dr. Debra Fischer, Astrophysicist, Yale

Future Astrophysicist _____ :

(Your Name Here)

To start down your path
Begin learning physics and math.
If math and physics simply aren't for you
There's no reason to be blue.
You have the strength to struggle through
As long as you believe in you.

Study the sky with astronomers in your city
You can find them in your local astronomy committee.
If a group cannot be found, there's no need to worry
You can create one in quite a hurry.
You'll need a few friends and some parents too
And a blanket and dark sky to enjoy the view.
If dark skies are rare, you can read books and go online
NASA's website is an astronomy shrine.

When you're ready for college, you'll need to choose a degree
You can pick from math, physics, and astronomy.
Manage your time wisely and think about what you learn
And good grades will be easier to earn.
Be honest with yourself about what you do and don't know
Questions and reflections will help you learn and grow.
Surround yourself with people that support your ambitions
People that see the greatness of your mission.

As you travel this path, there are some things you should know
Not everyone will understand your dream or think it can be so.
Your heart may hurt and your eyes may become teary
In these times remember: not everyone's thoughts are rooted in sound theories.
For this reason, grand and noble journeys such as yours
Are sometimes cluttered with closed doors.
But that's okay because you're smart and strong
So no door will stay closed for too long.
No matter how hard things may seem
You'll never give up on yourself and your dreams.

One day you'll look back on your time in college
And you'll be proud of yourself for gaining so much knowledge.
You'll realize that you've become the person you wanted to be
That against all odds, you created your own reality.
Others will recognize your efforts too
And you'll become a role model for children just like you.

Jupiter and its volcanic moon Io

About the Author

In a world full of stereotypes, Autumn struggled to overcome hers. She thought she could do things that others said she couldn't, she felt she belonged in environments that others tried to keep her out of, and she didn't have any role models in her fields of interest. The Science Starters Collection is Autumn's attempt to help children overcome their struggles by exposing them to the futures they can obtain and providing them with role models they can relate to.

Autumn's interest in science started when she was just 4 years old. Her insatiable curiosity and urgent need to understand reality were fulfilled by the method of science and the breadth of its findings. Autumn's lifelong quest for knowledge gave rise to her passion for teaching, and writing allows her to learn and teach simultaneously.

Before becoming a writer, Autumn obtained a formal education in biochemistry and women and gender studies while working in the pharmaceutical industry. In her late twenties she entered the biotechnology industry and academia.

Today, she assists in the development of green chemistry experiments for the chemistry teaching labs at the University of Oregon. In her free time, she attends various science classes, writes The Science Starters Collection, and explores the awe-inspiring state of Oregon.

About the Illustrator

Meg is a comics and story artist based in the Pacific Northwest. On rainy Oregon days, you can find her bundled up, sipping tea and working on a variety of projects ranging from comics and graphic novels to storyboards and children's books. Aside from art, Meg loves reading up on archeology, history, psychology, scientific discoveries, and feminism.

She would like to thank her wonderful partner, Carson, for all his support through this project.

She would also like to remind the young people reading this book that they are in charge of their future, and with enough hard work, can pursue any passion to the fullest extent.

Image Credits

Cover image:
Milky Way; NASA/JPL-Caltech/University of Wisconsin; Spitzer Space Telescope
ARP 274; NASA/ESA; Hubble Space Telescope
NGC 6872; NASA/ESA; Hubble Space Telescope
NGC 6611; NASA/ESA; Hubble Space Telescope
Cosmic Caterpillar; NASA/ESA; Hubble Space Telescope
Antennae Galaxies; NASA/ESA/Hubble; Hubble Space Telescope
NGC 6302; NASA, ESA and the Hubble SM4 ERO Team; Hubble Space Telescope
Andromeda Galaxy; NASA/JPL/California Institute of Technology; Spitzer Space Telescope
M33; NASA/JPL/California Institute of Technology; Spitzer Space Telescope

Dedication:
Carina Nebula; NASA/ESA; Hubble Space Telescope

Table of Contents:
Andromeda; NASA; Swift Telescope

Pages 1-2:
DEM L316A; NASA/ESA; Hubble Space Telescope

Pages 3-4:
NGC 4833; NASA/ESA; Hubble Space Telescope
Planet Earth; NASA/NOAA/GOES Project; GOES-East Satellite

Pages 7-8:
Planet Earth; NASA/NOAA/GOES Project; GOES-East Satellite
Sun; NASA/SDO; Solar Dynamics Observatory

Pages 9-10:
Sun; NASA/SDO; Solar Dynamics Observatory
Planets; NASA; Various sources

Pages 11-12:
Sun; NASA/SDO; Solar Dynamics Observatory
NGC 4833; NASA/ESA; Hubble Space Telescope

Pages 17-18:
Westerlund 2; NASA/ESA/Hubble Heritage Team; Hubble Space Telescope

Pages 19-20:
Ultra Deep Field; NASA/ESA; Hubble Space Telescope
Galaxy images; NASA; Various Sources

Pages 21-22:
ARP 273; NASA/ESA/Hubble Heritage Team; Hubble Space Telescope
Ultra Deep Field; NASA/ESA; Hubble Space Telescope
Galaxy images; NASA; Various Sources

Pages 23-24:
NGC 6984; ESA/NASA/Hubble; Hubble Space Telescope
Earth-Like Planets; NASA/Ames/JPL-Caltech; Artist renditions
Full moon; NASA; Photograph from Apollo 11
Jupiter Moons; Jupiter & Io; NASA/Johns Hopkins University Applied Physics Laboratory/Southwest Research Institute/Goddard Space Flight Center
Sun; NASA/SDO; Solar Dynamics Observatory

Pages 25-26:
Ultra Deep Field; NASA/ESA; Hubble Space Telescope
Galaxy images; NASA; Various Sources

Pages 27-28:
Young Stars; NASA/JPL-Caltech/Harvard-Smithsonian CfA; Spitzer Space Telescope
Milky Way; NASA/JPL-Caltech/University of Wisconsin; Spitzer Space Telescope
Dumbbell Nebula; NASA/JPL-Caltech/Harvard/Smithsonian; Spitzer Telescope
Ant Nebula; NASA/Space Telescope Science Institute; Hubble Telescope
Helix 7293; NASA/JPL-Caltech; NASA GALEX
Crab Nebula; NASA/ESA/JPL/Arizona State University/Hubble Space Telescope
Eskimo Nebula; NASA, ESA, Andrew Fruchter and the ERO team (STScI); Hubble
Helix Nebula; NASA/ESA/ C.R. O'Dell/M. Meixner and P. McCullough; Hubble
NGC 2818; NASA/ESA/Hubble Heritage Team; Hubble Space Telescope

Pages 29-30:
N159; NASA/ESA; Hubble Space Telescope
Primordal Quasar; NASA/ESA/ESO/Wolfram Freudling et al. (STECF)

Pages 33-34:
Hole in Sun; NASA/AIA

Pages 35-36:
Hubble over Earth; JSC; Space Shuttle Columbia
Hubble; STS-125 Shuttle Atlantis Crewmember

Pages 37-38:
IRAS 12196-6300; NASA/ESA/Hubble Telescope

Pages 39-40:
Jupiter & Io; NASA/Johns Hopkins University Applied Physics Laboratory/Southwest Research Institute/Goddard Space Flight Center

www.ingramcontent.com/pod-product-compliance
Lightning Source LLC
Chambersburg PA
CBHW040345060426
42445CB00029B/3